This project began on the 27 January 2021

Movement Force of the People

(MFP)

www.mwanakin.com

Mwanandeke Kindembo

i

I want to take this precious time to thank my sister **Fitina N. Kindembo,** who helped me in editing some parts of this book. She has always been there, and she is the kind of a sister that every author would wish to have in life.

Resistance

To

Intolerance

By

Mwanandeke Kindembo

Table of Contents

Introduction

After taking a while to meditate upon the motto of our generation. I came to the ultimate conclusion that our minds are mainly focused on wealth and health. Even though I have mentioned these type of desires in my previous books, I believe that it would not be agonising to clear the reader's mind once more. The main types of desires that each individual must go through at some stage of life are: Love, wealth, health, power and knowledge. This is if you take into consideration that wisdom is reached when that knowledge is put into practice.

Therefore, our generation is more focused on only acquiring wealth and health. Love is not seen as something of great status. Power is left untouched by the hunger for power and knowledge is neglected. This is simply because our youth are putting more emphasis on physical rather than mental development. We are now living in the generation of body-builders, rather than walking amongst of those who are ready, or well prepared to conquer their daily desires. Furthermore, the people of today are incapable of accomplishing most of their dreams because they are living according to the liking of the opposite gender. Boys wish to please young ladies by showing off a strong physical body and women are aiming to please their fellow men by delving excessively in make-up. Although it appears to be a win-win situation to the majority of the people, it is in fact the sellers among the people who are truly benefiting from this useless "game of competition", which they have invented. The consumer has been given no choice, but to keep buying and consuming non-stop. Now I ask, how can you expect to be capable of controlling your emotions after you have become used to the feelings of wanting more for yourself? Or, in addition, how can you limit the desires of pleasing and attracting those who are around you?

Therefore, if a lady dislike a man with a beard, eavesdropping men will react by shaving off their beards. This, however, will result in them resembling women. Perhaps, they are trying to imitate their wives and compete for first place in the chase for beauty. This is why most of them are physically beautiful, or handsome, but lack the intellect to conduct and manoeuvre through their daily lives. They find that they are constantly trapped in the pointless game of attraction and success. We have more sweet words coming from the so-called motivators and inspirers. However, we can see no results from their teachings except that we're only making them rich. These motivators are no better or worse than pastors, or any religious leaders out there. They will drain your money from your pockets while you are wide awake and then leave you more broke than before. The results of a successful life will be seen through them, even though most of them started as disadvantaged people. No doubt that many have bought private-jets just by brainwashing the innocents minds of the young people! Furthermore, gymnasiums have become the holy houses for the young and old people alike in the current world. Six-packs is their "promised land"; the main goal that everyone wish to accomplish in the gym. Even these six-packs does not last longer without continuous and excessive training!

It is all beautiful, at first, to put health before everything, but you will come to realise that it is foolish if you base all of your life on this single desire. Wealth and health are opposite attributes, they have different destinations. For example, you cannot wish for a good life that is full of joy and peace without sacrificing something in return. If you long for exceptional health, then try to forget about gaining wealth at the same time. For those who wish for great worldly possessions, you must forget about having health and peace of mind too. They are like water and oil, they cannot, no matter how much you try, be mixed together, nor confused when you come near them.

Finally, the wealthy people must sacrifice all, including having sleeplessness nights, at least most of the things that you consider to be blessings, in order to reach to their "promised land". Apart from that, most of them are forced to live on medications and food supplements to help them see the next day. You must be ready to sacrifice your health if you wish to gain wealth. That is simply an unavoidable law of nature. It is the rule of those who wish to reach to the summit, in order to help and guide those who are still at the bottom of the ladder.

At the end of the day, it would be wise to turn all of your focus and energy on achieving mental over physical strength. The body is the slave of the mind and not the other way around. You will simply be wasting your energy by putting too much emphasis on the improvement of the body, while leaving the mind idle — leaving it to rot. Activate it and it will soon become your compass in life. It will become your torch to cut through the darkness and confusions that are hunting you daily. It will free you from those invisible chains that used to pull you back, and thus, guide you to the right path. The path of self-improvement and not that of working 9 to 5. Free yourself, just as the slave seek for full freedom from his/her master. No man should become a master of another. You must resist falling into this category. You are all born free and equal, and free you shall all die.

PREFACE

As the title proposes — resistance to intolerance — you can already predict that this book is going to be filled with great new ideas, that have the sole purpose of illuminating the mind, rather than merely arousing it. Sometimes, we resist to accept different views or break away from the kinds of nonsense that we hold dear to ourselves. For it is not forbidden of you to have your own opinions nor from considering them to be great above others. However, it would be deemed sinful of you to refuse to change even though you know that you are completely wrong. Perhaps, you've sensed that someone else is right and you have been proven wrong. But keep in mind that it would be an exaggeration to tell or convince another individual that they are COMPLETELY wrong. The truth, or rather, reality is far from any human can ever reach. Therefore, no matter how hard we try, we cannot reach the complete reality of anything without becoming mentally exhausted of ourselves. But there are some truths, or common-senses that does not need proof and which are universally accepted as truth whenever they are presented before your senses. These are the types of truths which cannot be denied by anyone whose mind is sound or still working correctly according to our surroundings.

Therefore, it would be wise to know and understand when you can resist something in life. Know the right time to break from your ignorance and join the ship of those who have chosen wisdom. Perhaps, you can stay on the shore and learn everything the hardest way; self-learning, is most advised than following others blindly. Become your own leader and master in life. Begin by mastering your emotions, then your mind. Cut away all of those feelings that showcase your weakness and develop the strong character that you have always been dreaming since birth. Extend your hand to take that which was meant to be yours all along. I urge you to live to your fullest potential and to never limit your thinking according to the appetite of another individual. Explore everything that surrounds you. By everything, I mean you should also include the things that you still consider to be evil before your senses. It is only when you taste them that you will finally understand their true nature. Do not read about them then choose to believe everything without researching yourself. You need to live according to the experiences that you have gained, or wish to gain throughout your life. The Creator placed you in this world for a purpose and that purpose is to learn from experience and not relying too much upon the hearsays. Again, break free from your chains and ropes. Remember, those who are teaching you to remain in one place explored everything around them. They lived to the fullest of their potentials. That is why they got something to teach you, simply, because they are talking from experience. They refused to follow the masses, nor be held down by the opinions of others. Even prophets, saints and the greatest geniuses are not free from our argument here. They are, or were humans just like you, walking under the same sun, moon and the sky. This is unless you wish to give them the same spiritual powers like that of the angels or demons.

Otherwise, you are no lesser nor greater than them. Only your mind will define you at the end of the journey. Perhaps, even the angels are laughing at our inability to explore the planet around us and are, instead, living according to the teachings of other beings. We end up creating more demons, multiplying chaos and destruction, adding more to the original number that descended from heaven. Thus, disbelieving in the One who created us and giving more power to demons and humans who rose above all the superstitions of this world. Show me the prophet, or any great person who did not possess a rebellious mind and spirit. If you can show me, with full proof, then I will have no choice but to withdraw my argument in the near future.

The Beginning of Religion

Regarding heaven, the majority of people are embarking on a journey in the name of finding a better place to enjoy the pleasures of life. However, none has ever set on a journey in order to have a peaceful death. For, death is regarded as an evil and our minds are only meditating upon life itself, and never upon the destroyer of pleasure, which is death. Simply, because it is impossible to keep imagining the outcomes of death while living a tough life at the same time. It, therefore, demands great courage and faith on the behalf of the believer.

Since the promised land of all the religions is the achievement, or going to heaven, it would make sense to set your mind upon great things that are similar to those that you keep imagining you will receive above. Heaven is filled with only unimaginable pleasures and promises the absence of miseries. Furthermore, you are not forced to turn this world into a hell, thinking that you will receive, or gain the passport of entering heaven. You can still conduct your life peacefully — even to the point of living in luxury — while at the same time setting your eyes to the greatest prize in the land of the Most High. Remember that, most of the prophets began as poor people, but they did not die in the same state as they began. Some of them indulged in great wealth and enjoyed all the kinds of pleasures that are found within this world. At the same time, imagine them then being given the highest place in heaven. Therefore, it would be wise to imitate their actions and focus less on their words, or teachings which many of them did not practise.

The majority of them preached against slavery, but we are, afterwards, shocked when we read the truth about their daily lives. They were, or are men, just like any man walking today. They had personal desires as well as goals to achieve in this world. They were the best planners and most of their dreams came to be realities, just as any man wishes to achieve all of his dreams in life. Actions can easily reveal the truth about the person, but also remember not to ignore the words. For we are following the religions today because of the Word that was left behind. Even though we did not witness nor see any of the previous prophets during their times of preaching, we still put full trust in their words and call ourselves believers. We are, indeed, following those sweet words left behind and not examining any of their actions. All we can rely upon are the hearsays, which are passed on words, from mouth to mouth.

I am still struggling to put the crown on either religions or science. Would the crown be one crafted from gold or of roses? Indeed, it would make sense to crown religion with roses and science with the crown of gold. The former will be filling with emotions and forget to thank that individual who picked those roses. Forgetting that he had to avoid the thorns and leave only the roses. Perhaps, his bruised hands were soaked in the blood, but no one cares as long as they can see the beautiful roses presented before them. The only sin will be if that individual narrate his full journey and the difficulty of picking the roses to the people. This is simply, because they will become emotionally dependent and feel sorry for the man who picked, and made that crown of roses. Meanwhile, the unharmed scientist king will be happy to wear the crown of gold on his head. Yet, at the end of the day, he will get tired of it and store it somewhere safe. Otherwise, the crown will drain his energy and he would not be able to stand up from that throne, nor be able to descend down to speak to the people, who are placed far away from him. He would prefer something lighter that he can put on or take off at will, while the real crown is left safely inside the house or perhaps, left to the person appointed to look after that crown of gold. On the other side of the coin, the roses will also gradually come to rot and thrown in the trash. Same goes for the gold which will come to be melted once again and turned into anything that will prove to be useful in the near future. Perhaps, what used to be a crown will transform into a golden sword that will shed blood, conquering or forcing other countries to kneel before the king. Thus, the roses will vanish with time, but the gold will be melted with the force and heat of fire.

I ask you to imagine that you are asked by a stranger: "What is the philosophy of religion?" The shortest answer will be: If you come up with a new name of God, or the Creator, then you will be labelled as the new prophet. Thus, creating your own imaginary heaven and becoming a great model that your followers will want to imitate throughout their lives. For many claimed to have seen the Divine One or to have ascended to heaven and come back. Perhaps, there are different heavens for each individual, therefore, we cannot refuse, nor are we in the position to deny their claims. Every enlightened person claims to have reached a divine place. The place that is filled with euphoria; happiness, or to the final destination for all humankind. Thus, we can easily claim that, your definition of heaven lies wherever your self-interest and happiness dwells.

Trust not these prophets of today, who are still using the same old Bible as a referencing point. A prophet is nothing without a new revelation. Come up with your own book of revelation, or the "Word of God". The Word equals the book or, holy scripture. That is religion. There must be a place where all the correct teachings are written. Therefore, you cannot put any blame on any religion, but upon the followers of that religion. For religion is the book and the words within it. It is innocent from any committed action(s). The Word, or Holy Scriptures are filled up with beautiful words of the Most High, as most of the religions claim. However, the majority of people are reading the holy scriptures and basing their lives upon them. The first challenge is that, you cannot expect everyone to get the same understanding of the book. Are we not all unique in life? What about those who cannot read nor write at all? Their faiths will be based only upon the hearsays of those who have read, or pretend to have read the scriptures fully. Apart from that, you can also expect different explanations from each individual who has read the scripture(s). It is all true that we have got the same senses, but our systems and outputs cannot all be the same, even though we received the same inputs, or message through our senses. For example, you and I are looking at the same painting or drawing yet, we will have completely different interpretations. This is simply, because we have different perceptions and form our own concepts according to what we have been through in our lives. Either we are judging from experience of life, or basing everything on beliefs. Only the few who have dedicated their energy and time shall come to understand the truest meanings of the scriptures at the end of the journey. Therefore, we cannot tell, at first sight, who is correct or wrong, because we are all caught up in a different perceptive, views and opinions.

Moreover, God is love but God has got nothing to do with religions. Same as love is free from any religious belief. It can be said that, the Creator is free from all the needs. He is the giver and you are the vessel, or the receiver. If God is all loving, then we must treat the heart as the throne of God. Meaning, religions must try their best to turn and imitate the qualities of love, but love does not need, nor depend upon religions. Love is the motto of all the religions, but Oneness of God is the motto and the main message of all the Abrahamic religions. Perhaps, you can claim that love is your religion thus, loving every creature of the Most High, without segregating anyone in the equation of humanity. Love came first before any religion in the world.

That is the difference between faith and beliefs. You can have faith without becoming religious, nor having any religion. But beliefs is when you follow everything that you are told, read and see others are practising daily. You can try to use your reasoning, but the majority of the people follow blindly and decide to go with the flow. Like a leaf carried away with the stream of water, it can crash along the way, or go on to the darkest parts of the unknown ocean. The leaf may then reach the land of confusion, where it will not be capable to save itself once it has reached there. Where you would not be capable of seeing what is beyond your sight, nor examining everything that is presented before your senses. You will have no time for examination, therefore, accepting everything without a second thought; unable to use critical mind to sort out the rotting fruits.

Furthermore, religion is not an unrealistic phenomenon. You must have faith and keep your daily prayer. We must keep reminding ourselves that, the heart is the throne of God and the heart, is the mother of all the desires. These desires can be fulfilled or suppressed only with full faith and not relying upon beliefs. Since love is the lowest type of desire, we must understand that we share this same attribute with all the living creatures, including animals. You are, therefore, required to examine both sides without any fear of going to hell nor fear of being cast out within your society. Open-mindedness is the ability to look at both sides without taking sides. Fanaticism must be put to death. Your religion is not the best nor the worst when compared to other beliefs. It is all great to preach to others about what you believe in, or your opinions about your religions, but you must be willing and well prepared to accept the opinions of others. Remember, you are not advertising your religion to anyone, but you are trying to clear the doubts between you. Thus, you will find yourself having no choice, but to learn the arts of tolerance and becoming less resistant to other people's views about your religion. You will begin to love one another, in the name of humanity. Without the desire to end the life of another individual just in the name of being right, or of practising the correct religion.

Any individual can claim divinity, or that he/she is a prophet of God. Or, make up some magical stories and relate them to God. At the same time, you must be ready to be called a 'madman' too, another name for genius in our generation. Be ready to be stoned too, similar to those who called and referred to themselves as prophets of God. On the other hand, Prophets are geniuses. Not of teaching some useless superstitions, but of conquering the hearts and bending the minds. You are told to explore and learning everything, at the same time, not to learn too much. Just in case you might end up finding the truth about your existence or truest passion in life.

In The Name of God

In reality, all religions only pray in the Name of God. Because no one truly knows His Name. Let's remind ourselves that there was no English prophet. So, who came up with the name "god" in the first place? What difference will the capital 'G' make in the place of writing the word 'god'? Why does the name 'devil' sometimes require a capital letter too? Is it to show respect? Or, it is meant to be done for every name, including the names of humans too? Since your name does not change its meaning whether it begins with a capital or small letter, why does the name God change its meaning when you begin with a small letter? Some individuals would rather spell it as "G-d", skipping the letter 'o' in the middle so as to make it silent, nor are they allowed to mention that sacred name in private as well as in the public. Why is it considered as a bad thing, or a sin to mention the name of your Creator?

Are you ashamed of the One who created you, or you are not placing Him first before everything else that you so love in life? We tend to mention the loved object or our Beloved One, more than anything that of little importance. But it is a different scenario in different societies and cultures. The Beloved is left last and the lesser important objects are placed first before anything else. That is life, perhaps, the start of religions. Most of them are based upon cultures rather than concrete faith. However, to be great you must divorce yourself from your society and culture. Nor can you mix your culture with your religion. You are required to divorce one and renew a proper marriage with the other. You must dedicate your full energy upon one single point if you truly wish to hit the target. Never rely too much upon luck, nor the famous saying of hitting two birds with one stone. Move away from imaginary thinking and embrace the reality that is daily staring at you.

On Tolerance

The concept of turning the other cheek to your enemy is far from practice. Why would you allow a total stranger to slap you twice, for no reason at all? Is this the correct definition of tolerance, or it is just an imaginary concept of heaven?

In reality, no human being, with the right mindset can ever allow someone to slap them while sitting idle. Unless you are chained down and you have got no choice but submit, without your will. Otherwise, you will be left with no choice, but to retaliate and seek freedom and that peace of mind. Even the one who taught that famous concept was not free from becoming furious, at countless times, to the people who went against his beliefs, or the teachings that he held dear to himself. Even the so-called Christian countries are still applying the policies of Islam and Judaism. Those which encourage the individual to embark on the path of self-defence, instead of letting someone slap you for no reason at all. Simply, because that is the reality or the truth of human nature. Otherwise, you will be violating the theory of the survival of nature. The theory of letting the fittest to stay alive while the weaklings to be taken advantage in life. Since no one wishes to become a weakling in the first place, it would make sense to turn to self-defence; so as not to show any type of weakness before your opponent.

On Religion being the Cause of Poverty

It can be said that about religions, that some prayers have contributed so much in teaching men how to complain and whine to God, instead of doing things themselves. Thus, they have become idle and expecting more than their abilities can ever achieve. We are turning so much inward forgetting to focus on the outside world too. We are becoming more spiritual, in our minds, but in reality, it is the opposite of spiritually that is practised by the mystics. Meaning, spirituality does not encourage idleness, but you must be more practical than those who are not spiritual. You are required to pray, meditate and practice whatever you have been teaching to others. That is the main road or the door of becoming a saint in life. Practice, rather than idleness. For idleness is the mother of all sins and not those imaginary actions, nor those scary images of the devil that you are holding inside your mind.

The majority of the religious people are expecting miracles to happen to them daily. Forgetting the famous motivation in the holy scriptures, that goes as "you will eat only from your sweat." No one else can help you except if you are willing to help yourself get out of your current situation. Every prayer must be accompanied with action, otherwise, you will be basing your whole life upon expectations. Upon the imaginary things that you have no proof, nor the certainty of attaining them after death.

Therefore, you cannot keep blaming the rich people for all of your miseries if you are not willing to act in the first place. Remember that, the correct definition of equality is that we came through the same door to the world. No one was meant to be poor in this world. But we must be ready to change our circumstances by changing our mindsets and sometimes, our environments.

Self-Knowledge

Once you know yourself, then you do not need a mediator between you and God. Nor are you in need of the prophet once you have found your way through. Same as you do not need your professors once you graduate from university. Every child must leave his parents to start his own family and become self-sufficient, without referring all the time at his/her parents. Perhaps, you can begin to apply that same concept when it comes to religions too. You must be ready to accept your responsibilities and never become too dependent upon any living being except the Most High. He must occupy your mind, heart, and soul. Remember Him, constantly in your mind and allow Him to become your final destination in this life and the next. Dedicate all your prayers to the Creator, nor are you to involve any individual at the time of dedication except the Most High alone. Every man must pay the price, or for his own sins and you cannot do anything to redeem their sins. Since the prophets have full confidence in the divine One, you must put full energy on the improvement of your soul, and not wasting time on kissing their hands and feet. The rest will be left to the believer to use his/her reasoning in order to understand the correct ways of obeying the Creator, rather than a man. Boldness is something that you cannot take away from the prophets and geniuses. But you are too scared to try that which will put your mind at ease. You wish to remain and be treated as a slave, even in religions. Know your worth and move on with your daily chores.

The main job of the prophets is to guide and re-kindle the lost souls from going astray. Apart from that, you should meditate and dedicate your entire life to the Creator. The task of the teacher has come to an end and it is the right time for the student to outshine the master. Perhaps, your teacher would like to see you do better in life than him, rather than limiting your reasoning according to his teachings.

On the other hand, we cannot deny the influence of religions upon the Renaissance, that began in Italy. The idea of a man is the creator of his destiny, or you can create and destroy at will. That was the time of self-realisation, self-improvement and the journey of tolerating different views of others. Buddhism came up with the famous concept of "Nirvana", Christianity offered us with the famous saying "The Kingdom of God is within you" and Islam come up with "Allah is closer to you than your jugular vein"; that is heaven for some people.

Those words re-kindled those sleeping flames within people, during the time of renaissance. It was the final call that man is the creator of his miseries and those calamities can be turned into benefits by simply, knowing your worth in life. By putting your mind into action, you can simply achieve anything that you keep craving for. The correct translations of those words are: You are part of that divine force. Your soul emerged from the source, which is the Creator in this case. You will easily understand them when you begin at the very beginning of the creation. You are not a god, but part of the Most High. But you are complete as a man or as a woman. In other words, you have been given everything that is required of you to live in this world. The rest is simply fulfilling your pleasures or indulging in the worldly pleasures that are placed for your senses.

The Enemy of Science

It is only the vulgar people who will imagine religion to be the main enemy of science. Even though there seems to be a lack of communication between religion and science, they still share so much in common. Simply, because only siblings are meant to fight. You can only quarrel with someone whom you share almost everything in common or a similar person. But you cannot start a fight with a stranger unless they have offended you in the first place. Because you are most likely to respect a stranger more than you respect your parents and siblings, the people whom you know inside out. Considering that, you have been living together with them since your birth. Otherwise, if you were separated at childhood, they will remain as strangers to you; although you are all related somehow.

Therefore, the main enemies of science are facts and common-sense. Scientists are meant to prove or disprove these facts, instead of inventing more and feeding them to the young generation. Science is not meant to base its teachings on memorisation of facts, but exploration and self-experiencing everything in life. On the other hand, our educational system is filled with the art of memorisation, how much you can recall in the exams hall, and nothing to do with the testing of our intelligence. Thus, judging the student on his/her performance of that single day of examination. In other words, passing or failing the exams according to his ability to retain the information at the required time. In the case of sports, the athlete must go through a series of tests to make sure they are in great condition of performing the following task. But during the examinations, the students are not tested for their mental and physical strengths before entering the exams hall, or class. Thus, it will be considered as a stroke of bad luck if you get sick on the exam day(s) and sometimes, lose marks for being absent during that crucial day.

Just as religious people are aiming to enter through heaven, the scientists have their heaven too. The heaven of the scientists lies in knowing how the Creator did everything. From the imitations seen in Artificial Intelligence, trying to create a robot that can work and function just as a human being. To the ignorant scientists who are coming up with their new theory of aliens, as the highest intelligent beings in the universe, in order to replace the concept of God. Dinosaurs also are meant to replace the story of Adam and Eve in the Garden of Eden. The beginning. Perhaps, they are too ashamed to awaken the story of the dragons and convincing people to believe in them. Therefore, we end up with the concept of, humans are to God as the robots are to Man. There will come a time that many of us will be forced to say 'thank you' to the robot(s) after performing a great task that was supposed to be carried by a human being. Perhaps, that credit must be given to the scientists, or rather, all the inventors out there. But most of the ignorant scientists are basing their beliefs on theories, it would be wise to forgive and allow them to have something to rest their imaginations upon. That will be called as freedom of thinking, rather than of speech. The aim of science was not meant to turn against every religious belief, but to illuminate the minds of the people. Forget these childish competitions and start performing your task as required. For, you can still be religious and be a scientist at the same time. That is why religion is not the enemy of science. But it is something that keeps humanity motivated to move forward and explore more. It is a win-win situation for those who are reasoning correctly.

Whether you call yourself as a scientist or a believer of the Most High, you are all equal under the judgement of the Creator. You are all enjoying the sunshine and the benefits that come within it. None of you can claim to have invented the sun, nor the moon, so why are you still fighting with each other? Why are you quarrelling like siblings who have never learned how to respect their elders? If you all claim to be open-minded individuals, why do you find it hard to comprehend and live with different views of others? What benefit are you gaining by hating your neighbour, simply, because he/she is religious, or a scientist? Do you imagine that you are the only being who has the complete knowledge of the universe, or you think that you can go outside of nature? Where is the proof that you keep talking about? Why are you not using your own examples and stop relying too much upon those of the old? Since it is almost impossible to become a pious person, without having any self-interest left within you. Thus, it is impossible to claim to have a PhD in science. It is not something that you can complete, but science is continuous. It is all over around you. You are required to be ready and always experimenting with the daily phenomenon.

Remember, a genius is born whenever he proves many facts wrong. Relying upon facts is claiming that you "know it all", while a genius must go beyond these facts and prove them on his own. He must go explore and refuse to be among the mass, who are always ready to go with the flow and come up with something that will leave you shocked. That is why it is not surprising to see a scientist with a doctorate, but they have never done anything great in life. Nor, are the religious people safe from this argument. You can be the religious leader, while you have not found that connection with the Creator, you still feel that emptiness within your heart and soul. Therefore, you must reconcile yourselves and put aside that ignorance which is blinding you from seeing the truth presented before your eyes. Teach and learn from one another, instead of fighting daily. Work as a team and save humanity from ignorance and intolerance.

The Relationship between God and Man

In order to understand the relationship between us and the Most High, we need first to decode the Holy Scriptures that are regularly presented before us. Thus, making is much easier to understand, or at least, we can learn the secret of religions and their truest impacts on our daily lives. It is said that all the holy scriptures are encoded in the highest language; that cannot be understood by the normal minds. But if we cannot even come to grasp their intended meanings, why were they written in such language in the first place? Were they meant for other living beings apart from humans? And why would the Creator create confusion, instead of illuminating our minds? Perhaps, the Creator is free from these scriptures and only people are dedicating, or putting words into His Mouth.

The truth must be told that, this language is not hidden, nor is it something that is far from us. It is, in fact, around us in our daily lives. The scriptures were intended for living beings, including humans. Thus, it will be foolish of the majority to limit their reasoning and, therefore, leave everything to the highest minds, or those who claim to have better understood them. Since religion is not a child's play, it has got the power to arouse the minds of all the people, young and old alike. It is the most powerful movement in the history of mankind. Thus, it cannot be neglected, nor put aside without further examinations. What if all the scriptures are describing a certain act in our daily lives that can be easily understood by the majority of the people? That is why the message was encoded in the first place so as to arouse the mind rather than revealing the truth to all. You must keep trying to reach to the promised land, or start from the very beginning of the creation of man if you truly wish to know the truth of life. Same as Alexander did not leave his crown to anyone, but left it to the "strongest" man. However, in the spiritual world, it is only the self-conscious individual that is considered as the greatest, and not the conqueror of the world. In reality, conquering temptation is the toughest task among them all.

Man and woman are equal before God and so are all the living things that are in the universe. As humans, we believe that we are the best and the most beloved of creation before the presence of the Creator. Perhaps, we are caught up in the feelings of selfishness and we cannot even examine anything that is far from the sight of our noses. Meaning, we have limited our reasoning upon ourselves. Forgetting the fact that God has created other beings before us, as it is mentioned in the holy scriptures.

The Creator will give His blessings to the human with the desire to fulfil his/her cravings. But those cravings of the human cannot be fulfilled, nor satisfied at one instance. It must be continuous without long delays, or gaps in between. Every prayer is due to a need or a lack of something. That is why the rich man does not ask for food, but for good health throughout the year. The Creator will then wait for a response from the human being. The response, or rather reaction, is when you decide to put in some effort to assure the Creator that you are engaged in the act of receiving more mercy from Him. Or, you still need more blessings from above. Therefore, you cannot expect to keep receiving without giving something back, or paying the price for every pleasure that you wish to enjoy. If you do not know that friendships imply give and take, then you are not worthy to be called a friend.

While you are putting in some effort, the Creator will then send more blessings in your path. Otherwise, the Creator cannot keep filling the vessel that is not interested, nor responding to the first blessing that He sent. Thus, your blessings will be equal to your engagement with your Creator. Action is required on both sides. You must become active and responsive. Otherwise, you will end up missing all the blessings that are bestowed down upon the whole creation. Like in every relationship, you cannot expect it to last longer when it is only one-sided. Friendship only exist during happy times as there lies a line of betrayal beyond those happy moments. That is what the majority of the people are doing, betraying their trust to the One who gave them so much pleasures in life. They only remember the Creator at the time of difficulties and not when they are still enjoying their lives.

The given vessel, or human, will then become satisfied with the joy and pleasures that the Most-High has bestowed. Thus, be ready to carry any commandment without fear nor complaint. Since you have accepted the gift of a Friend, you must be ready to do anything for His sake. You must be ready to be at His service at any given hour and place, without any excuses on your part. Accepting the blessings and beautiful gifts is the main road to signing a pact unknowingly. The more you expect, the more you become a slave and chained down by your desires of wishing more for yourself. When the vessel is full, you will know it without a doubt. This is not a one-day process, but is instead, continuous as time passes by. The more you pray and plead to Him, the more blessings you shall receive in turn and the more you are ready to be paid at the end of the day. Therefore, you must accept a gift only from those whom you wish to make as your masters. God is the best option to put your trust in, instead of relying on another human being, who cannot even keep their own promises to themselves.

Allow the fruit to fall and rot, in order to receive more. The seeds must rot before the plant can sprout for everyone to see. Make that to become your second source of receiving your blessings from the Creator. Whether it is directly or indirectly, both opportunities and blessings comes from Him in the first place. Be merciful for whatever you possess at the current moment and do not crave for that which is lacking in your life. At the same time, you must remember that you are blessed with the mind and have the right to achieve whatever you wish in this world. Thus, fulfilling your highest desires and keeping them from haunting you in the future. Success will then be visible in few days, weeks, or months. Or, after several years. Therefore, reaping whatever you sow for a very long period. The fruits will last longer or shorter, according to how you decided to treat them in your care. Use them wisely, in order to allow the rest of them to reach their fullest potential.

At the time, some people can bear more fruits than others depending on how you invested your time and how dedicated you were to your Beloved. God is love and He will always love His creations until to the very end of time; to those who truly believe in the concept of the last day. This is the revelation that comes directly from Him, the One who never sleeps, nor tiredness overtakes Him. He is the One who inspires all of us to write that which is beneficial to all his creations. He is the One who guides us to help humanity, through His blessings, mercy and permission.

Source of Enlightenment

Enlightenment is the result of accepting to receive the light from the Most High. Whatever comes down on Earth cannot be thrown up and expected to remain there, without any kind of force holding it up there. According to the laws of gravity, it will always fall down on Earth, down to the receiver. That is because the Creator only gives, but does not expect anything from us. He invests in us without expecting any interest in return. His ways are mysterious indeed. Because He gives to those who obey and even disobey Him. He does not choose nor discriminate any of His creation in the universe. Whenever you put that effort mentioned in the previous paragraph, you can expect or not expect for anything, but the blessings will still come in your direction. Thus, you must form, from the very beginning, the greatest abilities within you to see those blessings or opportunities in life. Without that ability, the opportunities will come and pass without you noticing them before your eyes.

Just as every vessel is not the same, therefore, enlightenment will also vary according to your patience, heart, mind, and body whenever it comes to endure challenges in life. Your spirit or attitude must be fearless in enjoying every part of worship and the companionship with the Creator. It is not meant to be unbearable, but it still requires some effort on your part to receive the seeds that will bear many fruits in the days, months, and years to come. Preparation is necessary for every task that you wish to undertake. The future will be brighter only to those who have prepared to receive, correctly, these blessings from the Most High.

The blessings, same as the fertile seeds, cannot be predicted from the early stages. Therefore, you must be satisfied with whatever the Creator has bestowed upon you. Nor are you meant to become jealous of those who have received more blessings than you. Remember, they have paid the price to receive those blessings. They have sacrificed, in private or public, some parts of themselves in order to reach that point in life. However, you are required to do your best, next time, if you truly wish to receive from the Kingdom of Heaven. It is the kind of Kingdom that is full of treasures and filled with joy emitting from each wall's corner. Everything that which is in that castle is glittering and shining by His blessings alone. All the possessions are patiently waiting to be given away to those who have passed the tests of life. Each vessel will be filled according to its capacity and how much it is prepared to bear at the end of the race.

Oh! human, what do you wish for? Do you want the shining diamonds and gold, or you wish to be given the seeds that will last and bear more fruits for you in the near future?!

Sacred Numbers

Take, therefore, these sacred numbers and keep them most dearer beside your heart. The numbers are **2, 7** and **9**. Truly, these numbers bear all the secrets of this world and that of heaven. They will unlock all the chains that are holding you back from entering into the sacred House of the Most High. You are all blessed and allowed to enjoy, celebrate, and enter through the dwellings of your Creator. You are given the power to create other beings that are similar to yourselves. It is required of you, therefore, to cherish them and teach your offspring the correct ways of the Creator.

You are all among the blessed ones, but you are given innocent eyes from seeing that which is staring back at you. The Most High is truly closer to you more than you can ever imagine. Dearer than a parent-child relationship. He must be your first and last lover in life and death. Your heart must seek and crave for His love towards you. Welcome Him beside you and your dwelling will be saved from ignorance and effeminise of any kind.

The Fear Within Us

Why do you fear that which is blessed and betray your own heart by calling it evil? Fear is the absence of reasoning. It evades us best whenever we become too emotional. It is a sensation that aid to arouse those negative thoughts inside our minds. Fear not! For the Creator will, sooner or later, open your innocent eyes to see that which you were too blind, or shy to see. You shall see that secret knowledge that is embedded in each one of us. Whenever enlightenment knocks on the door, you must allow it to come inside and then, embrace it into your arms. Accept it with two hands and never allow it to leave you with loneliness ever. The scriptures must be understood correctly and not reciting them only with excitement. We need to possess a clear knowledge of what they mean in the first place so as to clear all the doubts that are within them or inside our heads. Different experiences in life, will result in different perceptions of what we read, or hear from other people around us.

All of us have got the power to see and feel the divine residing within. However, this knowledge has been kept away from the masses, who are not ready to hear, nor experience it. Nor will there be any generation that will be fully prepared to receive the real meanings of the blessing and the ways of the Most High. His works are mysterious, but not hidden from our daily activities. See Him in everything that is presented before your senses and try to go beyond reason to understand their truest impacts in your life. All the living beings carry His Majestic Name daily. Even animals and plants are performing and carrying His commands. He is not the One who loves to be disobeyed so often, thus, He is the inventor of automation.

This truth can be seen in all the living things, including the universe itself. The planets are rotating and held with the force that is beyond our imagination. The sun is stationary for the reason that we cannot tell, nor explain to others. Day and night are nothing more than illusions before our eyes. The sun does not rise from East, nor set to West; it is all inside our minds and within the limits of our planet Earth. The sun does not move, but it is our planetary system that is rotating around it.

Why are you easily terrified then like little children who are left alone by their parents? Why are you not awaken enough to see His blessings that are surrounding you; and that which is within yourselves? Listen, whenever you are scared and having some senseless dreams at night; remember that someone else around the world is enjoying their lunch. While it is in the afternoon here in Europe and most parts in Africa, it is night-time in China. While you go to sleep, China is fully awake to perform her chores. Therefore, which East or West are you dedicating to the rise and the setting of the sun? It is rather, the East and West side of your country and not that of neighbouring countries. For, they too, have got their own East, West, North and South.

Language of the Universe

Hold fast to your Creator and seek His guidance in understanding the world in which you are currently living in. At the same time, never limit yourself from exploring the universe around you. Nature is everywhere and no individual can go outside of it. In fact, we are all part of nature. For, nature is the mother, the beginning, while we are only its children. The blessings of the Most High can be found everywhere you decide to turn your mind into use. For, we are nothing without His mercy and blessings.

Lower yourself down, in order to receive His endless mercy within you. Pray, meditate, and find a calming place to collect all your thoughts at the end of each day. If this proves to be a difficulty, then do so every two days, or at the end of each week. Three is an odd number, but number two is predictable. Therefore, it is usable for everyone. Even numbers are much greater for relaxation, while odd numbers are beautiful to doers and the lovers of adventures. That is the main reason to mix both, even and odd in our sacred numbers system. Number two is even. Whereas, number seven and nine are all odds. However, each number is carrying its meanings and have the same weight in power and miracles in the universe.

Meditate upon these numbers (**2, 7 & 9**) and the full message, or revelation shall be revealed unto you. Whether in the form of symbolism or in the form of enlightenment. Once you decode those numbers, you will not find any difficulty in interpreting any holy scripture presented before you. Thus, the famous saying of 'all the religions are the same' will become obvious before your mind; because you will come to understand that they have been saying the same thing throughout the age. For the message is the same, to all who have dedicated their valuable time to decode these "holy scriptures".

Religious Conflicts

The Western world will never understand the follies of the Eastern world, nor the East grasp the follies of their brothers and sisters in the West. These are follies because everyone believes their religion is the only true revelation from the Supreme One. No wonder most of the Northern and Southern part of the world are often excluded from these holy scriptures. In reality, this torch of religions has been shone from the East, it is the same as when we keep teaching our children that the sun rises from the East. The rest of the world is only adapting from the doctrines of the East. Almost all the known and famous prophets are those from the East only. The West and the rest of the world have embraced and turned these prophets into their own. That includes living, conducting their daily lives and freedom according to the teachings of these prophets — if "prophets" they may be called.

Therefore, instead of saying that "the sun rises from East", we can easily replace the sun with Christianity. It rose from the East and set in the West. Why are you fighting and causing endless quarrels then between you? What are the differences that you are yet too blind to see? What is so unbearable about each one of you that you do not seem to understand, nor accept? Set apart your absurdities and embrace each other without any hatred in your hearts. Reconcile between you, your brothers, and sisters.

Different languages and cultures have added more confusions to the understanding of these holy scriptures. They are indeed holy to each living human, including the dead, or the wandering souls out there.

Holy scriptures are holy to every living thing that has a beginning and end. Meditate upon them, not just for the sake of reading or memorising for pleasure. You must sacrifice pleasure in order to obtain the highest pleasure bestowed from your Creator. Just as at the start of a new business, you must spend money; to make more money lawfully.

Love is the motto, love is the beginning, love is the source and love is everything that your heart shall crave for, in this world and the next. Love and adore each other until you find that, respect will no longer be an issue between you and your friend or those whom you were previously calling strangers. For, you are all from one seed. You must learn to live in harmony and call one another brother and sister. Cherish one another and stand up against any sort of inhumanity in the society in which you are residing in. Let not any human treat you, or your fellow brother, like a slave. You are all born free and carry the noble spirit within you. Considering that, you are not identifying yourselves as the sack of blood, bones, and flesh. You ought to go beyond the corporeal things and explore the divinity that is residing within you.

The Portrait of God

Humans are the most inventive beings in the universe. They have invented the image that of a man and call him as their God. Simply, because they are humans possessed with the worldly desires, they also imagine their God to be similar, or at least, possess the same qualities as them. The majority of them have gotten tired of visualising an imaginary God and thus, created something that suits their own minds.

Moreover, since we believe that we shall all become spirits once we reach heaven, we are still making the mistake of holding an image of a corporeal God that is similar to our physical bodies. Perhaps, that was the first thing that made all the sprits revolt against humanity. Simply because we did not invent a god that was similar to their attributes. From that day on, we became selfish, loving our physical sides more than the spiritual. We became attached to the worldly possessions, as we were not capable of penetrating through the sky and peek at what was prepared for us. Some individuals decided to bring heaven to Earth by inventing the portraits of their god.

For God does not need a picture, nor any physical attributes that you may imagine inside your limited mind. All the pictures and imaginations of God should be inside your heart and soul. Learn the difference between faith and belief, then you will come to know and identify where your ignorance lies. You have been limiting and hanging your god on the wall, one side of the house, while believing that he's the most powerful being in the whole universe. Therefore, a white god will please the minds of the white people, while at the same time displeasing the minds of other races that differ from him. That is the kind of conflict or segregation you have been promoting within planet Earth. You have tried and sometimes succeeded to convince other races to accept your kind of god and reject their own. You have brainwashed their minds into believing that your race is the superior, while their race is inferior. Truly, they have all believed and submitted to your teachings. They have become authentic followers and the best imitators of whatever you planted inside their minds and imaginations since they were still young. You call that as the "portrait of God", or it should be called as the "promotion of God"? Simply, because it would be foolish to imagine that the Creator needs to be portrayed inside your dwellings. Perhaps, you wish to create the same ideology of the "Big brother", or being watched by that portrait instead of believing that the Creator is the All-Seeing and the All-Hearing of all your deeds and words.

Are you not ashamed that a little child will come over and cover that portrait of your god? Thus, becoming invisible from those staring eyes, or making your god blind to see that which is going on inside the room? You claim to be educated, but what kind of education did you receive? If you cannot even reason through the small things that are presented inside your own house? How can you allow such nonsense to control your imaginations? Or, how can you expect a statue to offer you food, if it cannot even save itself from the rain and cold? Examine now to which level you have been humiliating your god by displaying it in the public domain. See to what level you have allowed your mind to become limited to the corporeal things, instead of going beyond the physical body and mind. Perhaps, you wish to go to heaven and meet your physical god while you are a spiritual being.

Monarchies

Throw away that foolish idea of carrying aristocratic blood. For, all the drops of blood are the same in humans as in animals. Who deserves the honour and the name of "noble blood" between you and a lion? Who is the most courageous between you to undertake any challenge that faces you daily, or to be ready to sacrifice yourselves for the sake of many? Which king or queen will accept to die first before the so-called "servants" and "subjects". See now and then, to what cowardliness are displayed by these kings and queens throughout the age. They are the first ones to flee the country whenever troubles have befallen her.

They seek shelter somewhere else instead of defending that which they keep calling their own. Thus, the trophy of victory goes to the animals, mostly the lion, the king of the jungle, who does not show a hint of fear in the times of dangers. Victory is not for the so-called kings and queens, who are indulging in great pleasures by filling their stomachs daily; without considering those who are helping them to keep that kingdom in the first place. The hungry ones are praying and begging for daily bread, while food is thrown daily in the palaces. The rich man has got no time to waste in praying for food because it is not an issue to him. They are feeding upon what your sweat has earned and overlook your hard work. The poor man has sold his freedom of expression to the state.

For, you are not poor only in the lack of possessions, but also in your way of reasoning. All the types of pleasures will appear evil before your eyes; because you have allowed your body to adapt to the difficulty side of life. Pain has become your motto in life and heaven your final goal. All of us came from the same source, same tree, and the same Creator. Therefore, we must have everything in common except in ideas. What is the difference between daydreaming and imagination? Only the Creator knows!

Spiritual Journey

What is there then to be feared in this life? Train yourself from moving away from identifying yourself as the body. It is the main factor that is forcing you to live your whole life as a slave and a worker to meet other people's ends. Sit down and converse with yourself. Ask yourself why you are behaving in the manners that you believe you have been behaving around people. Why are you terrified in the physical presence of a person, but you feel more at ease when writing his/her name down on the paper?

If you have no fear of shouting nor writing down someone's name, then you have got nothing to fear in their presence. Halt and look straight at their souls, not the size of their bodies nor the weariness on their faces. Reconcile with all the types of fears and learn to forgive yourself in times of difficulty. Nor are you allowed to treat others as slaves. You must, therefore, know your limits of self-defence; so as not to cross the line of mercy and forgiveness. Truly, your Creator is the most loving and you must imitate that love too. He is the All-Seeing, you must imitate that tolerance too in seeing the best only in others. You must become a rational being that is capable to balance everything, big and small acts in life.

Physical and Spiritual World

In the physical world, we have got the souls with the bodies and in the spiritual world, we have souls without the bodies. Therefore, the only difference between the spiritual and physical world lies in having the bodies, but the souls are common in all beings. That is why we have some famous beliefs of imagining that our ancestors will become ghosts or simply, spirits. Thus, making it possible for them to see us, while we cannot see nor hear them. But some individuals claim to have spoken or sensed their presences. To some, it is just your imagination playing tricks upon you, or you been missing that person so much after their departure to the next world — if there is any dimension out there.

Moreover, we are all convinced that the devil was once in heaven and acted in the same manners as other angels. Perhaps, he was the prince of heaven and had so much more power than the other angels up there. Upon that, the question arises: Why would a devil, or any evil spirit need food if he was once an angel? For it is said in the holy scriptures that angels do not eat nor are they in need of any worldly possessions. The same question can be asked in the case of djinns, or jinn. How can djinns need worldly possessions if they are made of souls without bodies? Is it not food made to rest inside your stomach and nourishing the body in turn? Perhaps, even the spirits need nourishment, even though they are not corporeal beings.

Furthermore, how can you believe that a spirit; soul without a body, can affect a human being while they have no sense of touch? By human being here, I mean the bodily side and not the spirit within you. Since we have the bodies and the spirit, the whole fight must be spiritual or at least be dealt with emotionally. Simply, because they cannot touch us but affect our moods by playing around with our minds or when you allow your imagination to go wild and start thinking that there are demons inside you. If your soul is seating on its throne within the body, how come another soul may come to replace you in your kingdom? If we believe that it is possible for a human being to be possessed with a djinn or any evil spirit, then it would be wise to be well prepared and alter our mindsets according to the environment we have been living in. You must focus on the important things that matter in life, instead of wasting your valuable time upon some imaginary beings trying to possess your corporeal body.

Finally, every human being must die before going to heaven, that is, you must become a soul instead of carrying the heavy body with you to heaven. This will be the time for complete freedom as you will possess the same powers that are found within the spirits. On the other hand, why would a spirit wish to possess and reside inside your body while it is already made free in the first place? Why do we imagine that we are ready to enter heaven only once we become spirits or are dead? Are there not some beings that are already considered spirits and they are still living together with us down here on planet Earth? Are the demons, or any spiritual being capable of visiting heaven just as the angels can descend and ascend? What is it that is missing in between that will prepare us and facilitate our journey to heaven? How can a spirit be terrified or forced to cross upon a bridge when it can get on the other side without depending upon that bridge in the first place? Is the bridge the representation of that missing link, or it was meant to be a physical thing that is found within heaven? Why do we keep assuming that we will receive the physical things in heaven, while at the same time believing that we will no longer need our bodies? What is the purpose of throwing aside and leaving behind all the corporeal things down on Earth if your only wish is to receive the same things in heaven? Why are we lying to each other and pretending to have mastered all the mysteries of heaven, or that of the universe? In reality, we think we know, but we know nothing at all.

Gender Equality

Anyone who wishes to be proud must be proud only in your gender that you were born with without transgressing the gender of others. A man should not act like a woman, nor a woman should behave similarly to a man. Depending on what you believe in life, you are capable of achieving anything without changing your body somehow. For those who wish to transgress to another gender, they are truly the destroyers of this beautiful environment we are currently living in. These are the ones who are insecure, unable to believe in themselves and their genders. For, a man has nothing superior over the woman.

Unless your sight is limited to the corporeal bodies and not seeing any further from there. Purify your attitude, or spirit to be capable to receive that which the Creator destined for you. Do not follow other people blindly, nor heed to their slanders about what you truly believe in. Listen to their arguments, but pass without cursing any of them. Love should be your truest religion and humanity your priority in life. Allow your lips to utter only beautiful words. Purify your actions and words, so that you will not add more fuel to those who lack these attributes within them. Approach, teach and question them whenever they are in doubt or when you truly believe they need the blessings of the Most High. Love is the beginning. Love will allow us to restart from the very beginning.

The woman is the mother, sister, and daughter to the man. You must know that the woman is carrying the world on her back. Indeed, she is the backbone or the foundation of the universe. The woman must be kept as she is; a woman. The woman must feel secure in being a woman in the first place since the woman is the blueprint and the foundation of the house. The man must be the keeper or the shepherd of the flocks. Every strong household requires strong foundations to function well. Without a woman, there will be no hope in the near future. That is your heritage given by the Supreme One, your praises from this beautiful world. It is not, therefore, the job of a man to force anything bad to the opposite gender, but a woman has to know her worth and feel more confident than before, or compared to the past generations. Be confident in yourself, oh, mother of nature. Feed your children according to their needs and save them from these plagues that many of the people are still too blind to see. Oh, beloved mother! Accept your responsibilities and never leave the post assigned to you by your Creator. Be like the bravest soldiers in the battlefield, those who never leave their posts assigned by the general. Whether it is a spiritual, moral, or physical fight, you must not allow your children to suffer. Oh, mother nature.

Our shepherd, or the man, is not to be left behind too in this fight of restoring humanity to its original form. You must not abandon your post as the protector of your wife nor be careless in terms of defending the family. The shepherd must be always fully alert so as not to allow the wild dogs or any family of foxes take away one of your livestock. You require proper training and readiness in every situation in life no matter which circumstances are presented before you. You must allow your spirit to explore the elements beyond the senses and be prepared for any outcome.

A believer is similar to a restless soldier in an endless battle. In religion, you are fighting with an invisible enemy or perhaps, your mindset and the power of your imagination. Therefore, begin by leaving vain words behind you and aim for more actions. Leave words to those who wish to flatter men only to end up working under the services of their masters. You must become the master of the house by perfecting your method of dealing with responsibilities — not by using sweet words. For, a man is the one who is judged by his accomplishments and not by appearances alone.

The Offspring

Your children are your seeds. Therefore, take great care of them by giving them nourishments from the very beginning of their journey. Teach them, before allowing someone else to teach them the opposite deeds that are going against your correct teachings, nor those which are opposing the Most High. For, religion is not a form of destruction in itself, but a blessing to those who have faith within them. It cannot destroy or bless you without your permission in the first place.

To some, it is a curse, because they cannot reconcile, nor bear to accept their ignorance before the Supreme One. These people pretend to know everything, but what we see is the exact opposite of pure knowledge and reasoning. They are blinded by many years of ignorance, and only repentance, sincere repentance, can save them from expressing their hatred towards other living beings. We start growing whenever we become aware of our existence and surroundings. That is what we refer as 'adulthood'.

The Calendar

To the atheist, the world is filled with madness and crazy people who believe in superstitions that cannot even prove to be true. The atheist has got completely different views from the theist, but each one of them is regarding their opponent as the foolish one. It is a comedy of thoughts and childish arguments these people are indulging in. The former believe in science while the latter in religion. Thus, we come face to face with where religion and science lack communication. Simply, because they are, or have been, speaking different languages throughout the ages. Fanaticism has blinded both sides from seeing reason and tolerating one another.

Science, or perhaps, our scientists claim to have completely separated themselves from religion. But what is the truth and reality that lies behind their claims? Science claims that the Earth is around 4.5 billion years old, in reality, they are still believing in the religious claim, of the Earth being in the year 2021. No doubt that any theologian can disagree on basing that magical number upon the persona of an individual who is believed to have lived before us. Precisely, in the times of the reign of Julius in Rome. That is the first illusion or lack of reasoning we get from religious beliefs. Thus, the planet is ruled under different calendars, according to the country and religion in which you wish to practice. There are different places in the world with different calendars, same as there are different religions with different calendars. It is the fight of knowledge; who has got the most correct facts or which sect possesses more wisdom over the other? The concept of BC & AD is the most popular around the world. We must remember that it is based on the world of an individual whose existence we cannot prove. On the other hand, we have got people who base their calendar upon the sight of the moon, thus, almost ignoring or being critical of the existence of the sun. They omit the fact that the moon gets its light from the sun and that the moon is, therefore, dependent upon the sun. Now, how can you believe or follow something that depends on something else? If there is no sun, the moon should stop emitting its light. The sun is more masculine, while the moon is feminine. Yet, we are taught to seek the masculine attributes and not act like weaklings nor effeminate in our daily basis.

Therefore, I wish to break free from this belief of basing any calendar upon the moon, rather than the sun. After all, I am a man, I must follow similar attributes of those which are possessed by the sun. Take, for example, if we were living on the planet Saturn, which has got around 82 moons, 53 moons are confirmed and 29 are awaiting confirmation. Now, which moon will you base your calendar upon? Since the sun is our common factor in the solar system, it can be seen that all the planets needs sun, but not the moons. The planets are not dependent on the moon(s) but upon the sun. Thus, the moon is not necessary for the universal equation, but a secondary option. The sun is standing all alone and it is dependent upon itself in almost the same manners as we claim that the Creator is self-dependent. Which one is more important now, between the sun and the moon? Or, which one deserves more credit than another?

At the end of the day, we see scientists and their fellow atheists who are operating under the calendar of that country, or the environment in which they are living. These people are preaching, or teaching, the facts that they do not practise. For, science will be completely separated from religion on the day that scientists invent their own calendar(s). Thus, practising what they are teaching and not preaching these facts to the children without any proof. This imaginary calendar is possible only when the scientists make up their minds on the correct date and year of the planet Earth. If they are still not confident enough in their facts, then they should refrain, completely, from corrupting the minds of the youth. They must refrain from teaching what they do not practise in the first place, and abstain from believing the things which they do not have concrete proof in.

Furthermore, the ideas of the sun rises from the East and sets on West will stop, including the idea of day and night, will cease to exist except inside our imaginations. The truth is scarce and lies are filled in our schoolbooks. I called them 'lies' because that is far from the reality of the universe. We are limiting our minds and imaginations only in this world, but we are too blind from seeing and experiencing what is truly happening from the outside. The sun does not set, nor rises, the sun is fixed at one single point.

The Ignorant Scientist & Believer

Today's world remains just as the old generations of atheists were laughing at the religions for believing in superstitions. However, they too were wrong in their reasoning and false proofs of believing that, the Earth was the centre of the universe and not the sun. It can be seen that; an atheist is someone who decides to see the wrong things in the world and refuses to accept the other half of the truth. It is the journey to pessimism, and not of those who are truly open-minded people. Nor are you considered safe in believing that which you have never seen nor experienced in your life. We are advised, by some holy books, to examine and test everything presented to us before accepting their nature in life. Now, coming to the religious people, where did the concept of accepting something without questioning and denying others without proper examination come from? Are you not forbidden to deny and accept heavenly things that are commanded to you?

Perhaps, that is the biggest contraction that shall never be solved in our lives, nor separated from religious beliefs. Atheism too has become a religion, just as their brothers and sisters who are theists by hearts and minds. You are all believing and following a certain idea that you lack concrete proof whereupon. If it is not an idea, then you will find yourself believing in a certain man or the words of another individual without proof. Therefore, you are all into the same boat, but heading towards different destinations. The sailors or members of the crew must not fight while in the boat, otherwise, you will be thrown overboard. Wait and keep patiently waiting until you reach your final destination, to begin the endless fights that you have been dreaming about.

Remember that, although you are all brothers and sisters you are still mostly divided in the quest of the truth. It is the road that troubled even the prophets because none of them claimed to know the absolute time of the Final Hour or the Day of Judgement. That was the test to prove human shortcomings, the lack of the truest wisdom, or the truth in general terms. That knowledge is only under the command of the One who inspires men to write, preach religions and start new civilisations out of barbarism.

At the end of the day, it would be wise for the ignorant scientist to come up with his/her calendar and stop using the religious calendars and signing their dates under the religious ones. They must begin by proving to us that they are capable and confident enough in their theories, then show us their final calendar according to their studies of the planet, or the universe that they keep boasting about. If that proves to be impossible to create, then I hope they can try at least to become tolerant of the concepts that which they keep ridiculing daily in the public. They must practice what they believe, both in private and public, and show us how the true scientist, or atheist, must behave in this world. Show us by examples and not mere arguments that will lead nowhere.

Demonstrate to us, then we will believe and follow in the same path that they believe to be correct above all. The wise man must become a friend of the truth, not a fanatic of the concepts that are believed by many to be true only. Rip away from sinning against your soul. For indeed, your mind and soul will never find rest until the day you decide to feed them the truth. Experience above all.

Origin of Intelligence

We can understand the truest meaning of intelligence only after learning and understanding the paradox between matter and mind, or the meaning of brain and mind. These are two separate concepts that are still troubling most of the people, even during our age: the 21st century.

The main belief is that of separating emotions and mind perhaps, trying to put the mind above the emotions throughout our daily lives. Thus, avoiding being carried away by mere feelings whenever you want to get something done. Upon that raises a question: How can you sense your emotions without the presence of the mind? The most accessible reply is that emotions, or rather feelings, cannot be felt by an individual without the presence of the mind. Therefore, emotions are completely reliant upon the mind. However, all of them are forms of energies, or thoughts that are generated from the brain.

We have finally reached the road that leads to matter, that is the brain, which is part of the body, or an organ inside the body. Which is the nervous system, that transmits messages, or electrical signals from the brain to different parts of the body, arms, feet, eyes, etc. Allowing us to tell when the cup is full, or it is still half-full. It can be said that, without doubt, matter comes before the mind. Simply, because there is no mind without having the brain in the first place. However, we must use the mind over matter; to come to this conclusion. Meaning, to be capable of seeing the truest form of matter, we need the mind to guide us through this complicated journey. Perhaps, clearing our doubts, or removing all the illusions that are presented before our eyes. Moreover, it is overall impossible to penetrate through an object, by using our sense of sight alone, without the aid of the mind in the first place. Consequently, the matter can indeed be known to us only through the aid of the mind. The argument does not end there, since matter must be felt by our touching senses, it can be easily seen that it is different to that of the mind, which is invisible to the naked eye, nor can it be touched. Although, that does not limit the fact that mind is the result of the brain. There is no matter without the mind. Those thoughts or energies must be emitted from the brain. Matter itself is still complicated to understand, simply, because it is made of atoms and those atoms can be divided into sub-atoms, and divided much further into small molecules and so on. It is an endless loop. Whereas, the mind is unique to each individual and it cannot be divided into small parts. However, that does not mean we are limited to experiencing wild imaginations throughout the day. That is why we are constantly reminded to have thousands of thoughts in one single day. What does this mean? It implies that the mind is generating several ideas, and hence, thinking for itself. But what is the meaning of thought? Where do our thoughts come from? That my dear friend, we shall never know.

Perhaps, we will come to find out soon. Some individuals will claim that it is all originated from our inner feelings thus, taking this famous paradox much further.

The closing conclusion is that there is no mind without the brain. Matter comes before the mind when you put all the jokes aside. It can be seen that the brain comes first before the mind. Why? This is because the mind is the result/product of the brain. There is no mind without matter so, the brain is the mother and the mind is the daughter. Armed with that knowledge, we can easily diminish the idea of using the mind to understand matter in the first place. The daughter is the product of her mother, according to how she raised her throughout. The daughter needs proper training and care before taking the role of a mother too. Afterwards, she will, then, with proper training, become free and independent in her life.

That same paradox was raised in this famous question: "How do I know that my knowledge is knowledge?" In simpler words, how do I know that I have a brain or intelligence inside my head? You see, you need your mind to work it out and identify every organ found inside the body. Meaning, the brain will appear to be idle while the mind is being prioritised. But the brain is still the main engine that is making all the magic happen. When the brain shutdown, the mind cannot resist but return home to where it belongs. All the excessive imaginations will cease, and the mind will become completely trapped inside the brain. This example can be easily seen whenever someone passes away. You will be offered around seven seconds or minutes to recall all of your life at the time of death. Recalling as some people claim, seeing everything that you did in the past and seeing all your actions presented before you in the duration of that short time. That is the time when the brain is trying to recollect all the mind which has been going wild around the universe and try to bring it back to the current moment. This is to lock it completely forever or for it to be trapped inside the brain because the mind has originated from the brain since your birth. When you were still a child, you barely used your mind to your benefit. Perhaps, you only used half of it. Fortunately, the more you grew, the more connections you made throughout the journey. Now, the more you know, the more knowledge you have. The more knowledge you have, the more imagination. The more you can imagine, the more you will be capable to explore the universe and every living thing that is residing in your environment.

Therefore, as the human being is evolving with time, the mind and brain are also expanding or increasing in size with time. Even though matter, which is the brain, appears to be idle inside your head, it is still the main source, or engine that which is controlling the car. When the engine dies, the car too cannot be expected to move from its stationary state. A quick example can be seen in the case of telepathic, or telepathy; when you leave your body. Or, when the soul leaves the body to explore different dimensions unknown to the corporal body. Even though we have the djinns, devils or whatever you refer to them, they have got the souls without the body. But every human being has got the body with the soul within. And this body is the matter, the source of all the confusions. No matter how far you wish to roam outside the body with your soul, but that same soul will still come back to its original state if you wish to remain alive. Same as the mind. It doesn't matter how much you use your imagination, how many thoughts you can consider, perhaps you even have a thousand thoughts a day. In the end, whenever you go to sleep it will come back home like the prodigal son; coming back home where it was originated. Perhaps, sleeping is like hitting a reset button of thoughts, thus, generating fresh ideas and getting a sense of relaxation at the end.

Animals and Humans

I do not understand why animals will not be judged for their actions. Why most of the people are against the idea of having animals' heaven, same as us humans. If animals can love and be taken away emotionally. If the animals can kill for love just as some humans do in every generation. How come those humans will face judgement on the final day, while animals will become sinless? Perhaps, we ought to redefine the meaning of the word "sin". What is it and where did it originate?

Upon Demons & Devils

If angels can turn into devils, is it possible for demons to become angels too? What was the transformation from being an angel to becoming a hatred devil? Why flatter yourself by saying that you hate demons when they were once considered to be angels? Are you going to hate all the angels who are going to turn into demons in the future? And are you going to despise your own son who is considered as a devil by his companions? Who, may I ask, taught you to hate or love without your consent? What is there to love or hate between the angel and demon? Considered that, they were once living together and joyfully in heaven.

Demons inside You

Same as the angels can transform into demons, I cannot see the reason why humans or animals should not be considered to possess some demonic qualities too. We are all the creations of the Most High. We are no better nor worse than the angels. The angels too fail upon the test of perfection. The test of pleasure that every living being must go through at some stage in life. Because if they cannot even resist the temptations of human beings. It then shows that they cannot be considered as pure beings. Being tempted also shows the existence of free-will since the devil chose to become so, instead of obeying the commands of His Creator. It is well known that he got many supporters whom we refer to as "demons". Similarly, humans have free-will and can choose to either indulge in acts of good or evil. In short, there appears to be a touch of free-will in angels as in demons — same as humans and animals. The Creator has blessed all His creations with the freedom of choice. Perhaps, there is a time, or instance that every angel must taste that freedom at some age in life. Otherwise, it will prove impossible for the angels to distinguish between good and evil. Furthermore, It will be so foolish of us, to imagine that they are not talking from experience, but from hearsays. That same knowledge of good and evil will take away their purity and innocence. Thus, the angels and demons will be no better, nor worst, than humans and animals. What will be the point of creation without leaving the freedom of choice to that being in the first place? Even automated machines do not work 24/7 nor allow them to work nonstop for a year long.

What is free-will? There is no free-will, or choice, without considering about self-interest. It can be easily said that she is the daughter and the offspring of self-interest. This is the main force that drives many people to love heaven or hate hell. It is the quest of happiness and satisfying our self-interests in this life on Earth and the desired next life; heaven. Therefore, you should not blame any outside force, nor negative energy for your evil deeds. You are responsible for your actions and sins, in the same manner as you believe that you are responsible for your own life. Without taking that full responsibility, you can be considered the furthest from being an adult. Your free-will must be placed under total dominance or surveillance instead of allowing it to roam freely and untamed. We must not, at any cost, behave like little children who are ready to take credit for good deeds but run away wherever evil ones appear. Take full responsibility if you are still willing to walk with your chest pumped up while calling yourself as an 'adult'. All the inspirations come from God, but where did the devil get his inspiration to disobey the commands of His Creator? It was all inside his mind, under the domain of his free-will, or freedom of choice?

The same thing will be asked of you when you begin to blame all your sins on the devils. They will not be responsible for the actions and decisions that you made while you were still alive and enjoying every second of your life. You must learn how to pick the roses and the thorns. If you wish to humour yourself in pleasure, then you must be ready to bear the fruits of pain that comes afterwards. Your input will be estimated to be equal to the output that which you wish to gain at the end of the journey. More efforts mean more gains, and so on. That is the universal law of nature.

On Judgement Day, you will be left on your own, tirelessly defending yourself according to what you did while still living under the sun. This is the most crucial part of being in heaven; you are not allowed to bring in any negative force, nor accuser to help you get ahead in your defence. The devil will be innocent on that day except carrying his own sins, that he committed from the very beginning of the creation of a man. Apart from that, he will not help any creature, including humans, by carrying up your sins upon himself. For, his sins have reached the saturated level and they are enough for him. Since you claim that he is the cursed one, how can you possibly expect him to accept, or carry your sins upon himself on the last day? Why do you wish to add more sins to the creature; devil that has reached the final stage of sinning? Who would force the devil to tell the truth in heaven if he began by disobeying the commands of His Creator there? Why are you getting involved in a fight that you lack knowledge of its beginning and end? Perhaps, you are using the famous concept of, the enemy of your friend is your enemy too. Are we supposed to fight the devil? If it is all true, then whom are we helping in this case? Does God need our help in this endless fight that has no beginning nor end? How can we help, or how do we know that we have been helping heaven win the fight? You see, we are caught up in a great mess of which we lack the background information. It is the invisible fight that we are trying to get ourselves involved in. The Creator blessed us with the five known senses, to the point that we cannot even comprehend what is in heaven, nor in hell. Thus, subtly forcing all the religious books to fall under the category of fiction. Simply, because they are filled with the stories that are obtained from dreams, visions, and imaginations. Apart from that, we can still find the stories of dragons and monsters, that are meant to arouse the mind rather than illuminating it.

You are all given the same powers that the angels and demons have. You have got nothing to fear except the ignorance of your intelligence or lack of knowledge. However, you are also blessed with a stomach to feast upon the worldly possessions. The power to create another life among yourselves and the ability to touch the corporeal thing that are surrounding you everywhere. You are blessed with the body to touch and feel one another. Perhaps, to become emotionally dependent upon the world and its atmospheric changes.

Charity and Donations

The idea of donating to the needy seem to be very appealing throughout generations. But the truth must be told, as this is not an easy task to perform, nor is it carried in the correct manners as it was supposed to be. In reality, even donation proves us wrong in the search of the final solution to end human miseries in life.

Simply, because we become too proud of ourselves while giving, or offering something to someone else. Within large groups of people, we cannot help but show off just how much we are capable to give. Even though we have got nothing left at home. We still give up everything, just in the name of giving while forgetting to help ourselves first. How can one refrain from being too proud of performing a kind act? We end up feeling pity and remorse on the shortcomings of others. This we call humanity. At the same time, we sense a lack of motivation from the beggar. In reality, it can be said that the beggar has got the spirit of a gypsy and they are not terrified by anything in life. They do not even fear the cold during the Wintertime, willing to continue waiting. Therefore, the proper way of giving charity is to give only to those who are truly in need. Offer food, rather than trying to show off with a large sum of money. Give food and see for yourself that the children and the needy are also able to enjoy themselves.

Unfortunately, most of the charities have turned into big organisations. Like every organisation, you can expect nothing good to come out of it except turning into capitalists themselves. The organisers will fill up their bellies and forget about the people, who need those possessions. The leaders will begin to indulge themselves in all the wealth, while the targeted people are starving and dying of hunger daily.

Hold out a helping hand, while, at the same time, making sure that your help is reaching the right people. Give without expecting any worldly reward here on Earth. Give to your fellow citizens and those whom you still consider strangers. Give them half of whatever you possess and finish by blessing them at the end. Do not expect blessings yourself, for you are not in great need like the person whom you've helped. Pray to the Most High to bless and guide them out of their miseries. Pray for their better future and yours too. Finally, remember to put in more effort to accomplish more than what you just gave out. Accomplish more only in the best manner and not by taking away from the needy. Earn a better life lawfully and according to your ability, or the rules of nature. For, nature is the best of the teachers of mankind. Because no one or nothing in this planet and the whole universe can be outside of nature.

Nature is everything and everything is within, or part of nature. Some individuals can claim that nature is a loose word, but in reality, you are part of nature. No one or any living thing can go outside of nature. You are just trapped inside a bubble; nature. You cannot define anything that is beyond your sight and imagination nor can you describe that which is found outside that bubble. Your eyes are filled with illusions that are preventing you from reaching reality, or the truth of everything. The truth of your existence. Sometimes, we think and believe that we know, but in reality, we know nothing. Look through nature and nature will stare back at you. For mankind is one and has emerged from the same tree. Since the tree is useless on its own without the support of the roots, you must help out your fellow friend to keep humanity alive throughout every generation. The more water you pour under that tree, the more nourishments the tree shall enjoy and reproduce more branches. These branches are your future children and the children of your children. Your great deeds will produce good fruits thereupon, for more generations to come.

The Literate and illiterate Beggar

The beggar who can write and read requires proper motivation than daily support from others. Since it is almost impossible to know how to write without knowing how to read in the first place, let the literate beggar know where he/she is going wrong in life. Show them the correct way by giving an example of how you achieved your success. Make sure that you show them the truest path and not to leave them half-way through the difficult journey. Indeed, this will prove to be the correct way of giving out more support than tossing coins to the beggar daily.

Perhaps, some of you are still confused upon the meaning of the word "motivation". Here is its definition: Since there is nothing new under the sun, motivation aims to confirm what you already know by simply, clearing your doubts. That is why you are advised to clear the doubts of the literate beggar instead of trying to invent some useless concepts which will lead to nowhere. Nor are you to fill them with mere words that will not feed them at the end of the day.

It is only the illiterate beggar who needs full support from you both in food and better quality of educational programmes. This will open their eyes and expand their intellects to see the realities of life. Show them as much as you have learned in the past, or if you cannot teach, then find an exceptional teacher that will guide them through that divine road of education.

Perhaps, you might start to doubt, thinking that it is impossible to persuade an adult out of their ignorance and miseries so it would be wise to teach their children instead. Thus, these same children should implement what they are taught and begin to teach the correct ways in their parents' home. Remember that, the children are the fruits or hopes for a better future. However, having more than you need usually results in turning the majority of the parents into the conditions of homelessness. You can have more children only according to your ability to feed them, and not depending upon a secondary source that cannot be relied upon during the difficult times.

Teaching and providing great services to these children will awaken those sleeping spirits, or successful attitudes within their parents to do better in life. Apart from that, these same educated children will remove all the worries, or at least, minimise all the guilts felt out by their parents. They will put a smile upon their parents and become the rich people of the next generation. They have been armed with that perpetual fire to achieve great things or almost anything in life. All their fears will vanish with proper education and no one will ever stand in their paths of reaching to the promised land of success. Therefore, it can be said that all they had been lacking was someone who would ignite that flame within them. Education will always come to their rescue.

Allow them to know and find themselves in life. That is the kind of motivational and inspirational talks you should have with them. It can be carried out on daily basis, or at the end of each week until the programme is over. Converse with them and do not forget to teach them how to become great leaders. Never let them to remain as followers. Remember, you are doing donation in the way of illuminating the minds of the needy. You are simply there to remove their ignorance and dissolve their miseries. Afterwards, knowledge and sooner or later wisdom will take care of them throughout their lives. It will put all of their calamities or daily challenges in the past tense of life. Their parents will sooner or later become free citizens just like everyone else who are proud to walk with their heads up in the street. Thus, helping humanity and saving her in general.

The Power of Love

Love is not an evil nor good thing, it is neutral. It depends entirely on the way you wish to program it and the result will show itself accordingly. If you made a grave mistake while programming, then you will only notice it when it is not running as required. It is, therefore, upon you as the programmer of your life, to pay more attention if you truly wish to succeed at the end of your journey. From birth to death, love is the motto of every living being. Even the devils love their own kind and, they cannot be evil to themselves. Therefore, evil deeds are those which are not appealing to your senses, nor satisfy your self-interest in any manner. Love is everywhere. Love is everything and the mother of true happiness in life.

But what would happen if the truth about love is revealed to you? I mean, what if you are told that love is the cause of all the wars and fighting in the world, perhaps, the whole universe? You will end up becoming a patriot and killing for the sake of the love you have for your country. You will not think twice before killing another being whenever you are aroused with the pleasures of love. The list is endless and only those who have mastered, or at least, are still learning how to master their animal desire are the ones who have learned the real truth about love. It is the same reasoning that forces many mystics from refraining completely from marriage. It is left to the bravest and the strongest who are ready to try new adventures in life. For love is beautiful only when you have mastered your emotions and mind. Otherwise, you will not be safe from acting foolishly, nor recalling your animal nature in every relationship.

Therefore, it can be said, without a doubt, that love is the cause of birth and death. You are living for the sake of love and your life is not free from wanting more love or pleasure to yourself. Love is good. Like all the good things, they must be loved by the majority in order to be considered as such. Otherwise, we will find ourselves knocking on the door of hate and start imagining that it is the opposite of love. What is hate and where did it come from? Perhaps, the discovery of hate was the beginning of hell. Thus, associating love with the Creator, or heaven, hate to hell or the devil.

Love is the primary cure of all the diseases. But no one knows the cure of love. The word 'love' is a vague word, but the majority of us claim to know so much about it. There are many types of loves. From God to man, down to the animals and all the living things within the universe. In reality, we are still confused whether we love with the heart or the mind. Always guessing and never coming to the final conclusion. Therefore, we should not neglect the contribution of the ears and the crucial part they play in making an individual be attracted to another individual. Hearing those sweet words will put the heart at ease and thus, finding peace of mind. Moreover, whenever the heart is in love, the mind becomes confused about what to do. Even concentration become a heavy burden.

Coming to science, the heart is nothing, but an organ that is there only to pump the blood around the body. No matter how many hearts you try to operate, you will never see love inside. All you will see is the blood and nothing else. And you are incapable of witnessing those emotions, inside your mind, that keeps you worried throughout the day and night. Even the mind is invisible. Both of them, mind, emotions and love, can be said to be more spiritual than relying too much upon our logical side. Thus, we cannot fully describe that which we cannot touch nor see. Simply, because we have eliminated the most useful senses into the equation of examination. That is why we must limit our reasoning here, so as not to come to that complex paradox of mind and matter once again.

Humanity in Action

We all need someone or a helping hand in times of need. We are not complete, nor are we able to defy the famous concept of no one is perfect in this world. In my experience, I came in contact with an elder woman in the shop. She was selling her merchandises, or goods, as usual, but I quickly noticed a box of donation beside the payment machine. The box was half-full of coins left by other strangers like me. I decided to ask before offering the left-over coins. She explained that it was the collection of a young lady who is fighting with cancer in Dublin. These words caught me with surprise; the idea of seeing an unrelated person who has decided to dedicate her life and time for the sake of the needy. At the end of our talk, she explained to me that her husband had passed away from cancer too. That is why she is trying her best, or anything necessary, to save those who are still fighting with that terrible disease. She is a total stranger to the sick person. Yet, living far apart, in a rural area of the country while that sick individual resides and lives in the capital.

It can be well seen that we are all the citizens of the universe. The world is a better place only to few, but in our case, we must try our best to include all the creatures of the Supreme One into the equation of humanity. We are one with the angels, even though you claim that you are just a mere mortal. Who is guaranteed to live forever except the Creator Himself?!

You can still reach the level of the angels by doing great deeds in your daily life. It is not too late, nor too early to help someone who is under your service. If you have got no one under your service, then seek them to their dwellings. Reach out a helping hand without any hidden agenda, or desire to achieve more in return. Do not wait for someone in your family to die before helping your fellow human being. Help out each other. Offer a helping hand even though you still believe that you are just another stranger in this world. Help without aiming to receive anything back in the meantime. Indeed, if you reach out to others, that is the main definition of humanity in a nutshell. Since you are a vessel, instead of constantly receiving, you may have decided to give out too, just like your Creator does. Bestow to others. For that is the work of the Most High to His creations.

Life and Death

Apart from all the things that you keep imagining to be expensive in this world, death and time come to be considered as the most expensive things in your life. Therefore, they are similar to the masters that you cannot disobey, nor run away from. Time is directly linked to death. Meaning that each passing day indicates the decay of your body. Since every being must follow the laws of nature and that of decay, it would be wise to keep considering that you will come to die one day, sooner or later — the hour is unknown to us.

The soul is considered immortal; it is free from death, and it cannot be affected by it. Therefore, we have death, life and pleasure as the original masters in life. These are the things that we cannot control, nor wish to possess in our lives. Even the so-called rich people in this world cannot buy any of those mentioned things above. For as life is free to all, so is death too. Pleasure is the promised land of every individual and the blessed fruit that each individual seeks at the end of their journey. Time is still too complicated for us to comprehend and we cannot define it using our wristwatches, nor by using the big clocks on the wall. It is all true that time heals everything. But even time itself needs treatment too. Simply, because it is all over the place. The world has no common time. Thus, it is not a strange thing that we still see so many misunderstandings in our societies. There is no such a thing as the universal time that is accepted by all. Each planet in the solar system has got its own different time and thus, different days, hours and minutes. Our planet is still not functioning at the same time. Different continents have different times on their clocks and even the countries of those continents have different times. Time cannot be mastered, just like the unpredictable wind that leaves us wondering which direction it is going to blow next.

All three: time, life and death are mysterious, and they will remain to be so until the day we will come to possess the key of all illusions. It is well known that the older you get, the wiser you become. But we are forgetting one attribute in the equation of life, that is death, the destroyer of all the pleasures in the world and the whole universe. Perhaps, we can turn to the law of conservation of energy that states, "energy can neither be created or destroyed, but only converted from one form of energy to another." Meaning, death or decay does not exist except inside our minds. Our physical bodies will become spiritual and so on.

After understanding this, what is the point of being terrified of death? Are you not in need of freedom? Not the worldly freedom, but that of the soul from the corporeal things. Some people call it heaven, and some will call it full freedom. Thus, we can be prepared to wish yet, expect nothing more or less in the next life. It is just a transformation of achieving total freedom from the desires of this world and becoming one with the Most High. That is heaven to those who think and use their minds wisely throughout the day. It is rather the liberation of possessions than that of wishing for more possessions in the upper dimension, or heaven. That is the freedom of the mystics and the philosophers, or any individual who wishes to embark in the ship of the wise. For abundant freedom will be achieved only when the body is destroyed, and the soul is placed first before everything else.

What is Happiness?

The most immediate definition of happiness is that it comes from achievements in life. The more you achieve, the happier you are likely to live. It is not, however, found only in the worldly possessions, but everywhere. It can come as the result of performing a great deed, in talking and getting to know strangers that have been living in your neighbourhood and so further.

That innocent and genuine smile of strangers will leave a great impression upon your soul and mind. The mind might become forgetful of great deeds, but the soul does not forget anything that it considers to be beautiful and good. Nor can you ever force your soul to love, or reconcile with a known evil deed. By feeling the sunshine, or pleasure, we end up forgetting all of our miseries, or pains in life. That is the beauty and nature of humankind.

Therefore, happiness is considered to be the final call of everything. Perhaps, the final destination for many living beings. Your mind may still be wondering why the demons and devils are included in this beautiful category of happiness. You are right in wondering why before proceeding any further. It is without a doubt that we sometimes picture the devil laughing every time we sin knowingly or unknowingly. Perhaps, if you are blessed with the capacity to smile, remember that the demons are also enjoying the outcomes of our mischiefs. That is why happiness is not granted only to one living being, or a single category of beings but it is there free to all. It is meant to be shared to all — even those you consider as strangers, aliens, or any living being residing within the universe. Your mind must go beyond the corporeal things, just as science is beginning to imitate during our generation. For your eyes are the seat of the mind. Clear eyesight will result in a bright mind and thus, having a clear logic to solve complicated challenges presented before you. The soul will be watching, monitoring all the actions and taking whatever message comes to it, meaning, the soul will respond according to your senses. This includes all those known and unknown senses out there or, those which are yet to be discovered in the near future. The eyes work directly with the mind. That is why they are placed on your head and not under your hands or feet. In the case of a blind person, you will easily notice something beautiful that is missing into their lives. Moreover, that does not give us any right, nor upper hand when dealing with them. We must all be grateful for the blessings that the Supreme One has bestowed upon us, at the time of birth. Remember, we are all blind when it comes to love, or at the time of choosing our life partners.

The Creator is the most knowing of that which our minds are limited to understand and solve in every generation. Be grateful for every outcome, those you consider to be good or evil, and happiness should become your motto in life. Happiness will seek for you only when you cease running away from the things that you call evil. Reconcile with your heart and mind then allow your soul to penetrate through the veil of all illusions. Seeking for that key of all illusions in life. Therefore, you must quit complaining about useless things that are not beneficial to your sense of reasoning. Self-realisation must become your final target in life and remember to treat others in the manner as you wish to be treated with respect. Not to quarrel, nor treat them similar to your siblings. Love and embrace others in your arms just as you used to love your parents when you were still young. Imitate the love of a mother to her child and nothing should go out of your view any longer.

Approach them in good manners and let them know that you care. Show them that their happiness matters more than all of their possessions, those worldly things they keep aiming in life. Since no person or any living thing can go outside of nature, it will be wise to become adaptive to different environments and seek to reclaim your citizenship of the universe. Never limit your mind and imaginations on this planet alone. Explore everything. Tell them that this world is too small for you and that your eyes are fixed upon the sun, and not the moon that is still in our range. Tell them that we are all one living body that cannot be separated from nature. We are made with the atoms of the things that we consider beautiful or those which we admire and hate on a daily basis.

Advise them to look and go beyond reason. Ignite their souls and let them see the hidden beam emitting at the end of the tunnel. Allow them to conclude that intuition unites everything while the intellect is meant to divide and work in duality at everything presented before you. Oh, look! Dear friend, have I not shown you the correct path to achieving happiness in life? Remember, not to fight those who still believe in dualism. For their minds are still to mature and rise from the base of reasoning. They have not yet tasted the truest form of happiness in life. They are still confused between good and evil, man and woman, brother and sister, etc. They are forgetting that all of these names fall under the category of a human, or person. It is only your reasoning, or intellect, that began to separate and seek for division, multiplication, addiction and subtraction in life. Remember, we are all one and one is the number that cannot be divided by another number, proving to you that you are only wasting your valuable time; it will give you out the same number that was divided by it. Proving that we are one before the Creator. One before those who have reached the highest stage of spiritualism. One before the mystics. One before those who have tapped into their intuitive abilities. Finally, we are one before the religion(s) that believes, practise and claim the Oneness of the Most High.

But we seem to be divided before that individual who lacks or still relying upon his/her intellect and reasoning alone. Now is the right time to open your eyes widely and take grasp of the opportunity that is presented before you. Your ability is ready to be put in practice. You lack nothing except clearing your doubts and illuminating or replacing that ignorance with knowledge. Become tolerant and resist any type of ignorance that is based on dogmas. It is the time to reach the promised land that is full of happiness, and a promising future for you and your offspring to come. Now is the time to take the key of all illusions in your hands and achieve pure happiness in life. And you are not advised to live according to the appetite of others without examining everything in your free time.

What is Temptation?

Is temptation similar to affection, or it just falls into the lowest type of desire, which is love? Truly, this is the time when our moral values are tested, or come to the very base of behaving like small children. Either by blaming all of our bad actions and reactions to the devil. Whenever you mention the word "temptation", everyone will turn religious and begin to imagine a strange being, mostly the devil, forcing them to do the opposite actions that they were not supposed to do at that moment. Are we not old enough to be unafraid of the masks and fearless at the time of adversity? Perhaps, you thought that you were being religious by daily imagining that you are closer to the presence of the demons but that is the complete opposite to those who truly wish to believe in the Most High. You are not qualified to call yourself a believer if your heart and mind are still troubled by the images and the paintings of the devil that you come across. You are still acting similar to the little child who is terrified by the person who wears a scary mask. You cannot be religious while at the same time giving the devil the priority, or picturing him having the same powers as that of your Creator. Without breaking away from your childish imaginations, it can be easily said that you are still religiously immature.

Strive to get away from that invisible fear of yours similarly to how you are willing to fight with the devil, the most famous enemy throughout the generations. In the past and generations to come if that fear is not destroyed inside the minds of the ignorant people, it may destroy them instead. Have you seen that which you are daily terrified of? Or, you just read about it somewhere and took everything seriously without a second thought? What right do you have in claiming and calling yourself as an "adult"? What are you afraid of then, if you cannot even describe all your fears, nor the things that produced that fear within yourself? Are you quite sure that the devil is capable of forcing the people to sin in Africa, same as in Europe, Asian and America, at the same time? If that is all true, does that means the devil is capable to be present in two, or more separate places at the same time? Who gave him that power to become omnipotent and omnipresent like his Maker? By what means is this power is being abused daily and what would the Creator gain by doing so? Why would the Most High give the devil, or part of the devil, similar powers, or perhaps, closer in appearance to Him? What was the pact between them, or at what price did the devil received his powers? At what end should he return those powers? What is the benefit of returning the powers that caused way too many troubles and calamities in the world? Is he waiting for the Judgement Day so that he can see those powers being taken back to its original place? Perhaps, he does not even believe in the Judgement Day. Why are humans beings becoming so weak in believing in the Most High, but find it so easy in following the concepts that cannot be proved by anyone?

Break free. Break away and accept the fact that you can be tempted, or not be tempted, according to your ability and the correct preparations to bear the attractive attributes of another individual. For, we admire and love those people who constantly tease us, but we are doubtful of those who show us too much love. That is the beginning and end of temptation. Fortunately, you can break away from those haunting feelings by daily meditation upon life and controlling the mind from acting like a wild animal. Moreover, it can be said that temptation is the heiress of pleasure and only the things that we consider to be beautiful in the first place can be capable to tempt us from experience them in turn. You are a thinking animal. The kind of animal that must understand and know when the correct timing is to indulge in pleasure, at the same time, learning how to refrain from rejoicing too much due to those pleasures. Otherwise, you will be easily carried away by the feelings of affections, nor can you break free once you have allowed your body and mind to get used to them. Simply, because that was the kind of training that you wished to nourish yourself in.

What if we are nourishing the desires of temptations daily by learning the laws of attractions? It can be easily seen that there will be no manifestation, nor attraction of anything without setting your mindset upon expectations. These laws of attractions are teaching us the opposite attributes, of expectations, that we seek to be freed from. Would you desire to live your whole life upon expectations? Or, would you wish to break away from the idea and desire of expecting everything that you did not work for, in the first place? Therefore, it can be said that the laws of attractions are simply the art of expectation. At the same time, that does not limit you from exploring and improving your life daily, while focusing on the present moment rather than excessively thinking about the future. Anticipation is the source of all the humans' miseries. It is the main cause of losing countless blessings or opportunities that are constantly present before our eyes, during the present moment.

Finally, it would be impossible for any individual to accept the feelings, or emotions of temptation without wishing to gain something great in return. Meaning, you wish to indulge in the ocean of pleasure, but you will end up drawing only a single drop of water. Temptation itself is mesmerising, thus, it will take you by surprise and you will end up losing the fight as a result. Simply, because you were not ready for it, nor are you capable of fighting against temptation nor are the angels whom you kept considering to be pure beings. Those angels who sinned, or could not fight against temptation, became known as demons to us. But humans will remain to be humans. And it matters not how easily you are tempted to go astray from the right path, or that road which is considered to be the correct one to take you to the promised land.

Human Judgement

Most of the laws, or rules are easily broken whenever there is no one watching or approaching us in a given environment. That is to say, by nature, the human being is not great at following nor abiding according to the laws in the society. There must be an individual to keep reminding them of what is supposed to be done and at what given time. At the same time, remembering not to take most of their freedom thus, living in great harmony and having effective-communication among you. Otherwise, you cannot expect everyone to act rightfully, nor responsible in every environment.

Since we share some desires in common with the animals, it can be easily said that man is also an animal. Perhaps, a thinking animal that can distinguish between moral and immoral acts. The only difference between a human and an animal is in having inspiration. Apart from that, we use our instincts and are capable of loving and hating so are animals too. Most of our decisions are based on experience. More experience on something equals quick decisions. Less knowledge means fearful judgements.

Furthermore, the married man has adapted to generosity and thus, breaks the rules of the famous saying "you must remove only the dishes that you ate in." This is what single people are missing in life. To be married, you must become tolerant and avoid most of the feelings of resisting in every situation. You must learn how to accept the emotional-side and opinions of your significant other. The truth is that: We are unhappy married and unhappy when unmarried. In marriage you must first endure, pity and then embrace. You must break away from all the rules of selfishness and make sharing as your new motto in marriage. That is why bold individuals are tamed whenever they get married and become forced to be kind in life. They will become cautious instead of being too adventurous in every situation. Your words will become limited and you must keep watching whatever you are saying, unless you wish them to backfire in the near future. Remember, some marriages are similar to being under investigation. Be patient and ready to answer any unexpected question that may be presented before you. This is when your explanation skill will be put to test. You are required to explain fully without skipping anything that was meant to be told. Here, you have entered in a different dimension. A different world that is unlike that of the single people. You are no longer alone, but with someone who is looking up to you and regularly examining your acts. Be gentle to them and keep the vows that you made from the very beginning of your marriage. Live up to whatever pact you signed up for!

The only disadvantage of life is to come in contact with someone who does not know when and how to put their self-interest behind the closed doors. For, all the sins are the daughters of self-interest. It is the driving energy that force many individuals to ruin their and other people's lives without knowing it. Self-interest must be sacrificed if you wish to keep your marriage alive, or to keep your significant one attracted to your attributes. For, you will be judged less according to your willingness to work and perform great as a team. You need to adapt to several environments. Learn the arts of effective-communication, so as to speak effectively with any human being you come in contact with. Expect less from them. Simply, because anticipation is the source of self-interest. There is no need to keep craving to possess all the worldly things. You must understand that you are not the only being living under the sun, nor are you the only one with a stomach to feast upon all the pleasures. Give something to others. Allow them to enjoy, at least, half of whatever you have been enjoying all along.

Finally, remember to blame no one, including yourself, whenever some calamities fall upon you. Be at peace and reconcile with everyone and everything around you. Truly, that is the correct path to self-knowledge, self-improvement, or the art of self-realisation in life. Only children are allowed to wish and blame those who are the most powerful, but in your case, you have already tasted the bitter fruits of life. You must learn from your previous mistakes and swear never to repeat them. Remember, this is the only time that you are allowed to swear. It is your secret. On top of that, you are discouraged from advertising it to others. It is your life, so live it according to your appetite and not according to others.

Summary

How come we, as humans, keep claiming that we are given the free-will, while the angels or any creature of God have none? Where did the chosen one complex originated and how can other beings live without the existence of free-will? The most shocking part is that, we are worst at making decisions even though the majority of the people still believe in having the freedom of choice in life. Moreover, we are told that we can make any plans and decisions, but the Creator must determine upon what is correct for us. He is responsible for making that final decision in life. Thus, making us to look very similar to children who are offered toys to play with while the adults take the real items. It is as if we are given to play with lowly types of decisions, but not the highest ones. Same as in life. You are offered a chance to live and enjoy yourself in this world, but sooner or later, death will come knocking on your door, forcing you to pay the price of the things that you didn't know how to utilise.

The chef has prepared everything on the table and your only job is to feast and indulge in pleasure. How can you imagine that the waiter, or the chef will allow you to leave the restaurant without paying your due? How do you expect to rejoice in pleasure without tasting pain at the end of the journey? Therefore, the price must be paid by any means necessary. That is where the concept of heaven and hell come into the arena of religious beliefs. These people understood that it is almost impossible to indulge in pleasure without taking the injuries of pain. Pleasure and pain are remarkably close to each other and it is impossible to wish for one without also receiving its opposite attribute. It can be seen that, we are not worthy of the so-called free-will, nor are we free to make that final decision without any limitations pressed upon us. It seems that, free-will is not free at all, as we cannot even identify nor understand our truest passions in life. The beauty of living is found in having different opinions, tastes and desires. We seem to be confused and daily guessing upon what is good and evil, holy and impiety, etc. Our visions and minds have become caught up in the enigmatic veil of illusions. Guessing has become part of our lives that we have even started guessing the times of our deaths. We are still ignorant of that which concerns the correct timing of departure from this world just as we came unknowingly. Death is part of nature, but even nature does not seem to be free from the laws of decay. That is, everything that is alive must come to die too. Or, tasting the bitter fruit that comes along with death — the destroyer of pleasure in the universe. The land of perfectionism can be easily reached by avoiding the major imperfections in life. Thus, we automatically become unfaithful by referring to another creature of God as an infidel. Therefore, God should be welcomed in the heart and not in the act of worship alone.

Book Description

In this book, the author, Mwanandeke Kindembo, has presented the reader with his most troubling and yet, appealing arguments on the topics of religion and science. As these arguments were not meant to offend anyone, nor discredit any religious belief out there, the author has decided to act as a mediator. Hence, he reveals to the reader that which we have been misinterpreting in our daily lives. He has cleared many doubts and doctrines that are found within religions as well as in science.

As this is a book written on the foundations of self-help, he mainly focuses on the improvement of the mind rather than on the physical aspects of the body. This book unveils his philosophical thoughts to the reader and, thus allows for anyone to understand his thinking in general terms. It is, therefore, filled with critical ideas that will give the reader a new insights on life. It is a guide for the reader to follow in order to attain the promised land of happiness. The rest shall be left to the reader to conclude on their part.

The End.

Movement Force of the People

(MFP)

www.mwanakin.com

www.ingramcontent.com/pod-product-compliance
Lightning Source LLC
Chambersburg PA
CBHW070403220526
45467CB00001B/464